S0-CDO-715

The Book of the Ninja

Book Of The Ninja

Published by Paladin Press, a division of Paladin Enterprises, Inc., P.O. Box 1307, Boulder, Colorado 80306.
ISBN 0-87364-207-4
Printed in the United States of America

MANDAMUS

OBEY THIS WARNING

This Training Manual of the Black Dragon Fighting Society is the personal property of,

M____

Of the City of________________________________

State of______________________________

The same being a rare and valuable Volume of the Ancient Wisdom and Celestial Fire, written and arranged in the manner of Self-Instruction and, in consequence of which, every person is thus warned and most earnestly cautioned NEVER TO STEAL the same nor distort, nor tamper with, the Teachings or Formulas given therein, for if you steal this Volume, or by intention cause perversion of the Formulas, Sacred Names and Seals contained therein, you will attract an Evil Spirit and other powerful influences WHICH WILL MOST SURELY DO YOU INJURY AND CAUSE YOU TO REGRET YOUR ACTIONS TO THE DAY OF YOUR DEATH.

Should you be so foolish or rash as to steal this Sacred Volume, which is the property of thy neighbor, before you have read and recieved this Mandamus, RETURN IT TO HIM AT ONCE AND NO HARM WILL BEFALL YOU, but if thee DARES to disobey this warning, WOE BE UNTO THEE, for the Great System and Order of Adepts and Master Lamas of India and China have their Astral and Occult Guards who can witness your every action, see and know every thought that goes through your brain, and they will not hesitate, FOR THEY KNOW TOO WELL, how to avenge thefts and tampering with their Occult Teachings. So BEWARE and OBEY THIS MANDAMUS. Otherwise you will never know Peace, Fortune, or Contentment afterwards.

TABLE OF CONTENTS

INTRODUCTION---

This manual is for the information, guidance, and use of all Black Dragon Fighting Society personel. For ease and convienience to the reader, the material is divided into three primary parts:

Basic Techniques- the initial duty of any martial art is to provide its students with a variety of combative techniques. Section One deals with twenty such fundemental methods of attack and defense.

Advanced Techniques- these "kata" (pre-arranged exercise) not only provide the opportunity to rehearse the basic techniques in combination, but also comprise the deadliest, most savage, and terrifying self-defense forms known to man.

Sword Forms- Section Three deals with the specific training forms of the "Chien" or Sword, and are designed with the warrior in mind. They are the first five forms taught to the members of the Hsiao Chien Do Fencing Fraternity, and are revealed here for the first time.

The martial arts are an extension of the mind and the individual karateka must be provided with specific training to acquire the basic knowledge and skills. In addition, he must be not only properly indoctrinated and motivated to take appropriate action on his own, but also mentally prepared to follow the orders of those designated as his leaders.

Then it follows that the objective of this manual is:

1. Train the student in the principles of martial art.
2. Enhance the student's confidence in his ability.
3. Familiarize the student with the techniques of both planning and executing martial arts skills.
4. Instill in the student a sense of pride and of self awareness.

BASICS

DOJO KUN---

"To Strive for the Perfection of Character,
"To Foster the Spirit or Effort,
"To Defend the Paths of Truth,
"To Honor the Principles of Etiquette,
"To Guard against Impetuous Courage."

The origin of the name "dojo" is found in Buddhism, and means, "meditation hall". The Hall, therefore, must be as clean as possible, and within it there should always be that solemn atomosphere which ought to prevail in every place of worship or of mental training.

The Dojo is used for Training, Exercising, and for Matches. Sometimes it is also used for lectures, or for the exchange of questions and answers.

The hall must be covered with "Tatami", or Judo-Mats, which are woven throughout with single stitches of hemp or linen string to provide additional strength. It must be surrounded by panels about shoulder high, and all nails, angles of pillars, and the like must be eliminated to prevent any danger of accident. Visitors to the Dojo are requested to observe such etiquettes as "No Smoking", Hats Off, Be Silent, and so forth. Such an accepted code, which is found in Japan's martial arts may sound strange to persons who are unfamiliar with them, but still they are respected as ever by the masters and students of the martial arts as one of the valuable "legacies of the Samurai".

The aim of the martial arts is to instill into the mind of man a spirit of respect for himself and his fellow man. To this end, the Five Principles of the Dojo Kun are directed.

There are many things which the novice must learn in addition to the teachings of his martial art. First, the Dojo is a place of culture and one must therefore compose oneself and behave seriously without talking idly or acting noisily. Both at the time of practice and during a match, one must maintain a good deportment and be attentive to others who are exercising, and by watching them, try to learn some lessons which

are helpful in improving oneself.
If one should wish to secure the best results from the practice of the martial arts, one must consistently observe moderation in eating, drinking and in sleeping.
One must also, as a matter of course, refrain from eating and drinking water during practice, as well as immediately after the exercises. In order to obtain sound sleep, one must always finish everything without danger of being disturbed. Keeping one's body clean and wearing neat costume is necessary not only for health, but also out of consideration for others. One should remember to pare one's nails, not to neglect mending one's costume, and to make oneself comfortable before beginning exercies. During the exercises, one should close the mouth and breath through the nose.

Finally, co-operation should be the ruling spirit to keep the Dojo well arranged and to maintain the order of the Hall, since it is the common house of all who use it.

RANKING---

White Belt- symbolizing that the student is pure, and not yet initiated into the secrets of the Art. This is the first rank of the student.

Yellow Belt- This is analogus to the rank of Corporal in that the student is expected to set an example for the class, and be prepared to demonstrate those techniques with which he is familiar.

Green Belt- This is the level of the Non-commissioned officer, in that the student is expected to lead the class and is responsible for carrying out those duties which he may be directed to perform by his senior belts.

Brown Belt- This is the first level of command. The holders of this rank act directly in the name of the ranking instructor, and are responsible for directing the training of the other students.

Black Belt- This is the level of command. Those who have attained it have done so with patience, perseverance, and practice. They are required to set an example not only for those who follow them, but also to the people beyond the Dojo. They are the leaders among men. In Dojo, they are addressed as Sensei, which means "teacher". They act at the direction of the Grand Master and Conscience.

As there are many styles and systems, there are also many "sub-divisions" in the above ranks. Brown Belts and below are measured in "kyu" which means "boy", their grade being determined on a declining scale, such that Brown Belt 1st Kyu would be a higher rank than Brown Belt 3rd Kyu. Above the rank of Black Belt, the grades are measured in "Dan", and indicate term of service. Here the degree is measured on the incline, such that Third Dan shall be of greater rank than First.

MANUAL OF ARMS---

TACO- (tah-co)Taco, or the Position of Attention, is the basic martial arts stance. It indicates that you are alert and ready for instruction. Assume this stance at the command LINE UP or TACO. Thereafter, move only as ordered until you are dismissed.

a. Bring your heels together smartly on the same line.
b. Turn your feet out equally forming an angle of 45°.
c. Your legs are straight without stiffening or locking the knees.
d. Hold your body erect with hips level, chest lifted, and your shoulders square and even.
e. Let your arms hang straight without stiffening along the sides, with the backs of your hands outward. Your fingers should be curled so that the tips of your thumbs touch the first joint of your forefingers. Keep the thumbs along the seam of the trouser's leg, with all fingers touching the leg.
f. Keep the head erect and hold it squarely to the front with your chin drawn in. Look straight to the front.
g. Rest the weight of your body on the heels and balls of the feet equally. Stand still and do not talk.

SESGAE- (sess-gay)This movement is made from the Taco position only. At the command SESGAE, move the left foot smoothly 10 inches to the left of your right foot. Place your hands behind the back, just below the belt line, both hands extended and joined by interlocking the thumbs so that the right hand is outward. The head and eyes are held as in the Taco position.

RIGHT or LEFT FACE- This is a two-count movement. The command is small letters is the "preparatory command" and the command in capital letters is the command of execution. This movement is performed only from the Taco position. At the command FACE of Right FACE, slightly raise the left heel and right toe; turn to the right 90° to complete count one. On the second count, place your left foot smartly beside your right foot in the Taco position.

ABOUT FACE- At the command FACE of About FACE in this two count movement, place the toe of your right foot approximately six inches to the rear and slightly to the left of your left heel to complete count one. On the second count, turn 180°, stopping with your body in the Taco position. Arms are kept alongside the body during the entire movement.

KNEEL- At the command KNEEL of Ready KNEEL, step back with the right foot approximately 18 inches, bending the right knee and placing it beside the left heel to complete count one. On the second count, place the left foot beside the right and lower the body to sit on the heels. The hands rest in the lap, the back is straight. The head and eyes are held as in the Taco position.

BOW- This is a courtesy which is rendered to every Black Belt and Brown Belt you may meet or see. This salutation is not merely an ordinary form of greeting, but is also an expression of respect for one's opponent of Sensei (teacher). It should be made seriously before and after exercises and sparring, since, by nature, these are forms of contest in which we express our state of mind; that we contend for perfection of technique and practice, so that the contestants must respect each other. There are two types of Bow in the Black Dragon Fighting Society, one is made kneeling, the other standing. The Bow must be directed not only at one's opponent, but also toward the dias or platform on which superiors and instructors may be seated, as well as to all those present, and when one enters or leaves the Dojo. The Bow is performed from the Taco position by placing the hands palm down on the thighs and inclining the body forward approximately 45°. The eyes remain always on the object of the Bow. In the matter of contests, this is to prevent the opponent from launching an unsuspected attack. In salutation, it is to observe that the Bow is returned. The junior belt always bows first.

MEDITATION---

Sit on a cushion or thick rug which is comfortable for you. Sit in the Lotus position if possible; otherwise sit in the Half-Lotus, or simply cross the legs. Discover the position which is most natural for the body, relax the shoulders, straighten the back, do not lean.

Close your eyes and empty your mind. This is very difficult because the mind is full of many things. Try to diminish your thoughts. This will help you see yourself inside.

Clench the fists and place them on the thighs close to the knees. Place the fists palms up. This will help to clear thoughts from the mind and enable you to concentrate on the experience of Inner Energy.

Put the tongue against the palate and inhale through the nostrils, drawing the air deep into the lungs. Hold the breath for a second and slowly release it through the mouth. This facilitates the removal of stale air from the lungs. The mind is centered at the point between the eyes and behind them. Repeat the breathing exercise twelve time.

The best time to practice is in the evening before going to bed. Morning meditation is also desirable after a night's sleep, when the mind and body are refreshed and alert.

The goal of meditation is the cultivation of the Force, the Inner Strength. To cultivate the Force within the body requires that the student first master the body, then understand it, then relax and become one with it. To accomplish these goals, meditation teaches us to use the Imagination.

Within this mystic diagram, simple as it may seem, are patterns and designs of every concievable direction- vertical, horizontal, or diagonal- that the hand or foot can travel whether they are used singularly, simultaneously, or in combinations. In it are circles, half-circles, quarter-circles, plus squares, rectangles, triangles, crosses, criss-crosses, patterns, diamonds, hearts, figure-eights, overlapping figure-eights, and spirals. The hands and feet may move concurrently or in opposition to each other, as well as at various speeds.

Kicks, pokes, chops, blocks, etc. can travel to any imaginable point on the compass. Whether the art is boxing, fencing, or wrestling; this pattern will advance all self-defense systems practiced throughout the world. However, it must be remembered that this pattern is only as good as the mind of the man studying it, and his ability to adapt it to his own system. If a man can use his Imagination, untold secrets can stem from the study of this diagram. It may not come in a week, or in a month, but by giving it constant thought, your mind will eventually yeild one or two principles that will be effectively advantageous to your present system. You will not find this diagram in other books on Karate. It is revealed here to the public for the second time.

ANATOMY---

A through knowledge of the human anatomy is an important factor in the study of the fighting arts. The vital areas, pressure points, nerve centers, major blood vessels, and accessible organs, are all vunerable to attack.

The following chart clearly illustrates the location of the major vital areas of the body which are considered "Primary Targets" for the body's natural weapons.

#1. Trachea (Windpipe)
#2. Lungs
#3. Blood Vessel
#4. Heart
#5. Diaphragm
#6. Stomach
#7. Upper Liver
#8. Lower Liver
#9. Large Intestine
#10.Small Intestine
#11.Bladder
#12.Testicles

A.Larynx; Esophogus; Carotid Sheath
B.Branchial Artery; (supplying blood to the arm)
C.Solar Plexus; (at the tip of the Xyphoid Process
D.Blood Vessel; (supplying blood to the Heart)
E.Heart; (above the Solar Plexus)
F.Upper lobes of the Liver; on the right, the Spleen
G.Lower lobes of the Liver; on the right, the Kidneys
H."Pit" of the Stomach; (striking upward)
I."Hara"; (two inches below the navel)
J.Urinary Bladder; Testicles

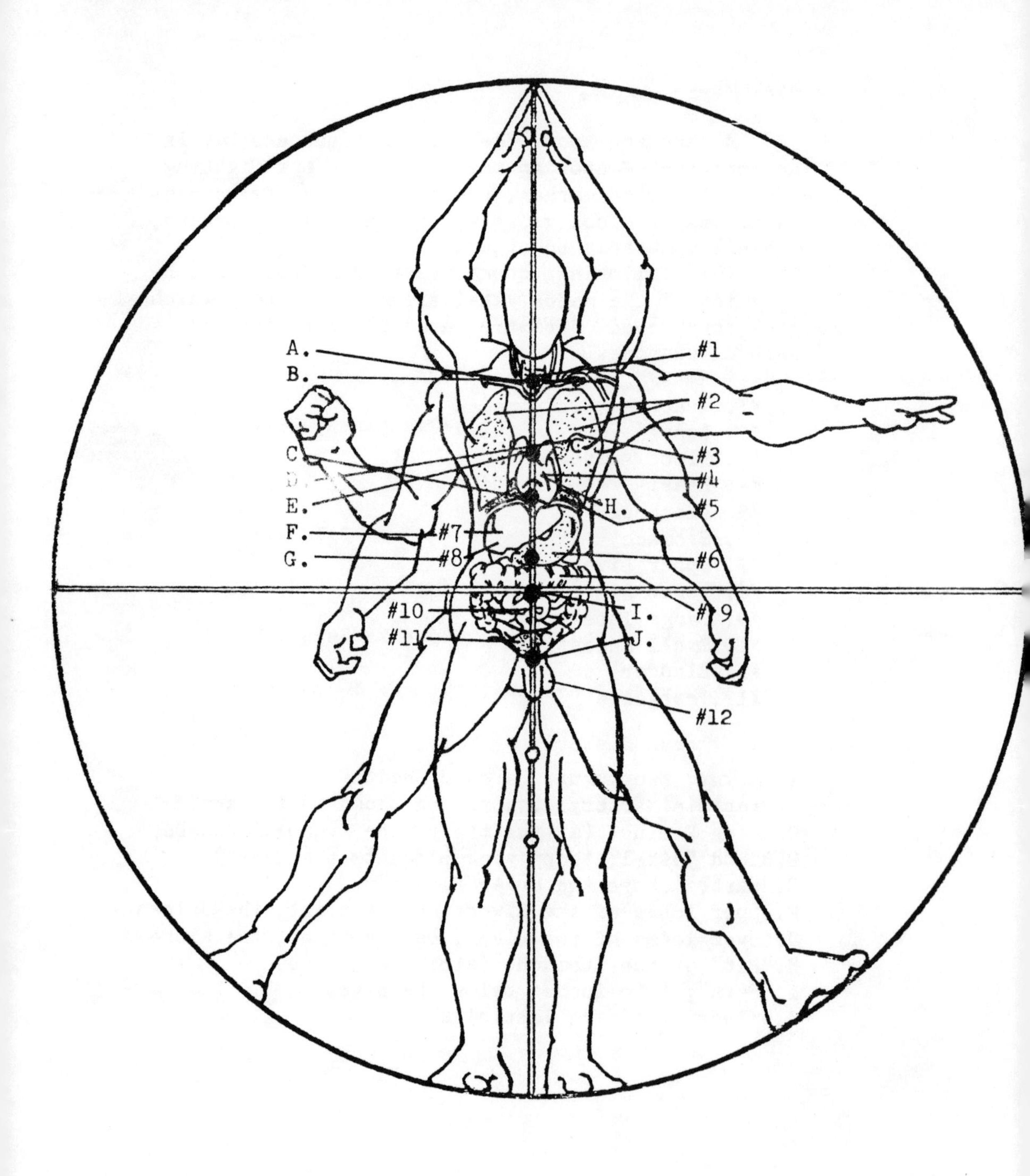
A.
B.
C.
D.
E.
F.
G.
H.
I.
J.
#1
#2
#3
#4
#5
#6
#7
#8
#9
#10
#11
#12

BROTHERHOOD OATH

We, are proud to be Karateka...

We, shall always pratice and study...

We, shall always be quick to seize opportunity...

We, shall always practice patience...

We, shall always keep the fighting spirit of Karate.

We shall always block Soft and strike Hard...

We shall always believe that nothing is impossible...

We shall always discard the bad...

We shall keep the good...

We shall always be loyal to ourselves, our art, and our country.

SELF-DEFENSE TECHNIQUES

STRATEGIC CONSIDERATIONS---

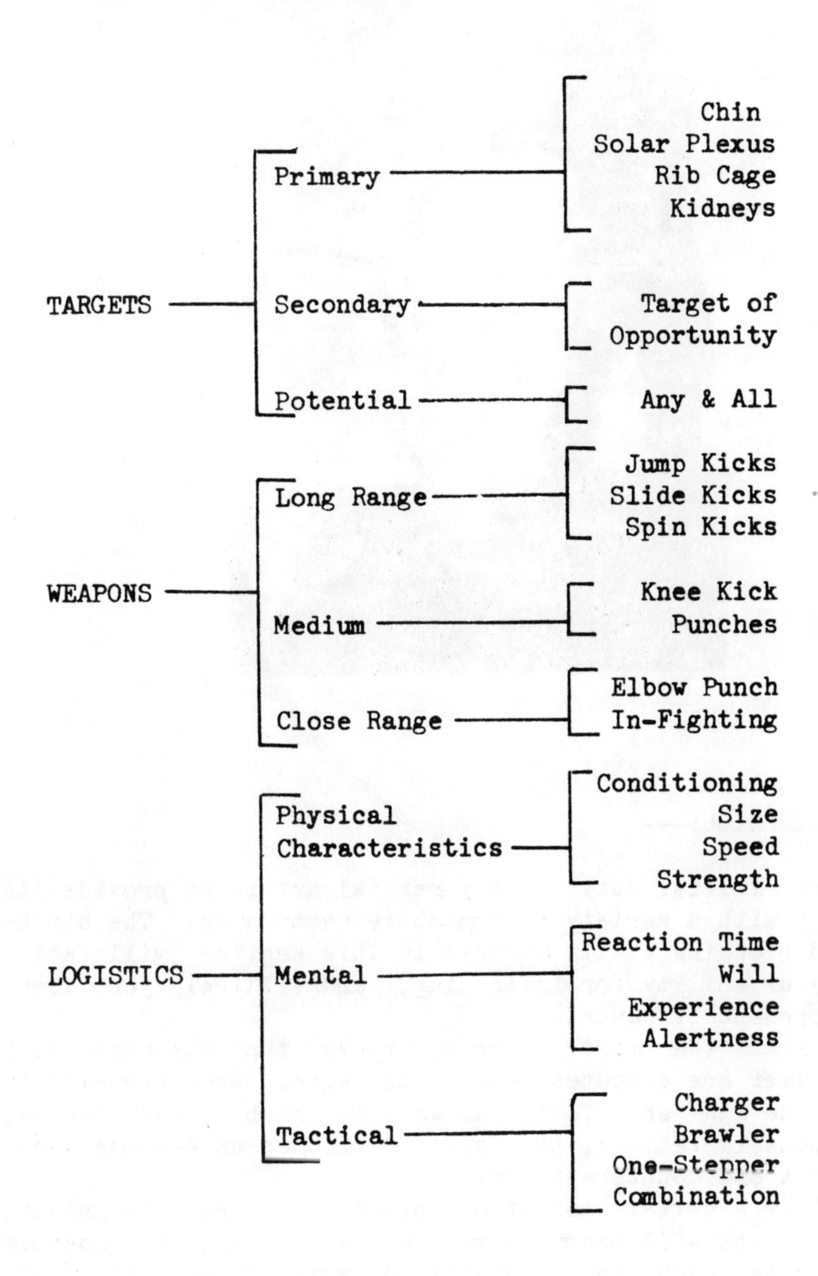

BASIC TECHNIQUE---

The initial duty of any martial art is to provide its students with a variety of combative techniques. The striking and blocking skills offered in this section will facilitate your ability for-infighting, and will help you keep your opponent off-guard.

Perhaps the most powerful blows that the human body can deliver are executed with the kick. When the ability to use the leg and foot, as well as the hand and forearm, as weapons is refined, the system becomes an awesome means of attack and counter-attack.

After a certain amount of practice on each technique, the movements will become more fluid and it will be possible to execute each in a single dynamic motion. Only then, should you proceed with the more advanced exercises.

HORSE STANCE---

Separate the feet two shoulders-width apart. The soles should be parallel to each other and the feet should point straight ahead. The knees should be relaxed, not stiffly locked. The head, neck, and trunk should follow a straight line, piercing the midpoint of the soles, excluding the heels. This applies the body weight to the broadest part of the feet. The body should be loose, the shoulders down, as relaxed as is naturally possible. The hands hang beside the hips, fists clenched, palms upward. One should breathe naturally, and look straight ahead.

KNIFE-HAND (Shuto)

This striking point is located along the outer edge of your hand between the base of your small finger and the bone protruding from your hand adjacent to the wrist joint. To isometrically from the Knife-Hand, flex this area by bringing your fingers tightly together at the knuckles and curling your thumb inward. Assume the basic fighting position and raise your rear hand to a position just above the shoulder, even with the head. Strike with a circular motion, and be sure to follow through in order to get maximum power.

REVERSE PUNCH (Seiken)

Begin by assuming the basic fighting position. Throw a punch at the Solar Plexus with the rear hand, remembering to twist it in a "corkscrew" action. The other hand should be brought back to your waist into the "chambered" position simultaneously. At the completion of the strike, your arm should be perpendicular and the shoulder should not be extended beyond the neck. The power of this strike comes from the speed with shich it is delivered and from the slight twisting of the hips. The striking surface of the fist is between the first and second knuckles.

BACK-FIST (Uraken)

From the basic fighting position, bend your forward arm in at the elbow and strike outward with it, hitting the target with the back of the first two knuckles. When the strike is completed, you should be leaning slightly toward the target and the arm should be extended but not straight.

RIDGE-HAND (Yoko Yubi)

From the basic fighting position, extend the fingers and form "knife-hands" by locking the thumb. Lean forward on the front leg and extend the rear arm to the side. Strike in a circular and horizontal motion, using the inner edge of the hand as the striking surface. Follow the strike with the shoulders and hips to insure maximum power, and aim at the temple or side of the neck.

FINGER JAB (Yubi Uke)

Begin by executing a shoulder block with the rear hand. As soon as the incoming attack is deflected, jab forward with the index and middle fingers toward the eyes of the opponent before he can withdraw his attack and cover his centerline. Follow through until the opponent is subdued and blinded.

CROSS BLOCK (Osu Juji)

From the basic fighting position, step outward and to one side with the forward leg, simultaneously pushing across your chest with the lead hand. This action deflects the incoming punch toward the centerline of the opponent's body, and may be used "high", as shown, or "low", against a kicking attack.

RISING FOREARM BLOCK (Jodan)

Begin by assuming the basic fighting position. Twist your hips toward your forward leg and move the lead hand upward at the shoulder and outward at the elbow. At the completion of the forearm deflection, your blocking arm should be perpendicular to the floor and your fist should be even with the top of the forehead. The forearm is the most effective part of the body for blocking.

SHOULDER BLOCK (Chudan)

Begin by assuming the basic fighting position. Lean backward on your rear leg and move your forward arm in a sweeping-circular motion in front of your torso. At the completion of the block, your fist must be even with your eyes and perpendicular to the floor and your shoulder.

DOWN BLOCK (Gedan)

From the basic fighting position, step outward and to the side with your front leg, and block outward and downward with your leading forearm. Perform the block in a circular motion, finishing with your arm extended and your fist pointing down and away from the body at a 45° angle.

FOOT-SWEEP (De Ashi Barai)

Assume a position facing your opponent. Seize him by both lapels firmly, taking a short step forward with your left foot. Push the opponent toward his right rear. He will resist toward his left front, but before he can do so, he must have taken most of the weight off his left leg. Thus, as he resists your push, you execute a low crescent kick, sweeping his foot away from the centerline. The combination of these two actions, plus a slight pull with your right hand, will upset the balance of your opponent in such a manner that he will fall on his side, perpendicular to your position.

SIDE-KICK (Yoko-Geri)

Begin by assuming the basic fighting stance. Lean straight up on your rear leg and raise your rear leg by bending the knee. As your rear knee becomes level with your waist, immediately pivot your anchor foot to the outside and allow the hips to turn into the move. Lean your upper body backward with your shoulders perpendicular to the ground while you simultaneously kick your lifted leg out sideways and strike the target with the "blade" of your foot. This area extends from the heel up to, but not including, the ball of the foot. Your foot is "curled" inward toward the inside ankle in order to expose this edge for kicking.

SNAP-KICK (Mae-Geri)

Begin by assuming the basic fighting stance. Lean forward on the front leg, bringing the rear leg up to waist level by bending the knee, then "snap" the foot out, striking the target squarely with the ball of the foot. When the kick is completed, your leg should follow the identical path back to its starting position.

ROUNDHOUSE KICK (Mawashi-Geri)

From the basic fighting position, lean straight up on your front leg and raise your rear leg by bending the knee. Continue this action until the rear knee becomes level with the chest while you simultaneously pivot the anchor foot 180 to the outside. Lean your upper body slightly backward and, in a circular motion, bring the rear foot around to hit the target with the instep or ball of the foot.

CRESCENT KICK (Harai-Geri)

Begin by assuming the basic fighting position. Lean forward on the front leg and bring the rear foot forward and outward in a circular motion. The striking surface is the inside arch of your rear foot. In the instance illustrated, the target is the opponent's kidney.

INSIDE CRESCENT KICK (Uchi Harai-Geri)

Begin by assuming the basic fighting position. Lean forward over the front leg and swing the rear leg forward as in the Crescent Kick. Allow the momentum of the rear foot to carry the extended rear leg up to chest level, and whip it back toward its original position with a twisting of the hips in a circular motion to strike the target with the outside edge of the foot. The foot should be vertical to the ground when the strike is effected.

TOP WRISTLOCK TAKEDOWN (Seio Nage)

Should the opponent seize your wrist, immediately counter by grasping his with your free hand. Continuing the momentum of this action, lift his arm above your head while pivoting to the right rear. Step behind the opponent with your right foot, simultaneously pulling his wrist over his shoulder. Step out and to the rear your left foot, while your right heel blocks his right ankle, and take him down. Care must be taken when practicing this technique lest you dislocate the opponent's shoulder.

HIP THROW (Ogoshi)

Seize the opponent by both lapels, assuming a position facing him directly in so doing. Step in and across the opponent's path with the right leg, thus preventing him from taking a step forward by blocking his right ankle. Grip the left lapel still more firmly and slide the right arm around the opponent's waist. Step to the rear with the left foot to block his left ankle. Twist the hips, flexing the knees, and lift him onto the right hip. Complete the throw by pulling down with the left hand. The opponent will land on his back in front of you if the throw is properly executed.

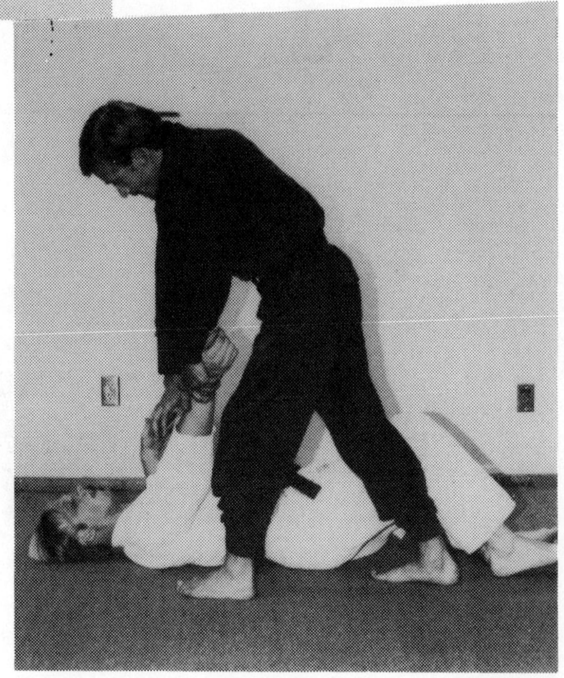

INVERTED THROW (Sukui)

Assume a position facing the opponent. Step quickly forward with the left foot, taking a long step, and placing the left foot behind the right heel of the opponent. Swing the left arm upward as you "step aside" to strike the back of his knee with your right wrist. This action will lift his foot clear of the floor, and the action of the left forearm will carry the opponent backward over the left knee. He will land on his face perpendicular to your position if the throw is properly executed.

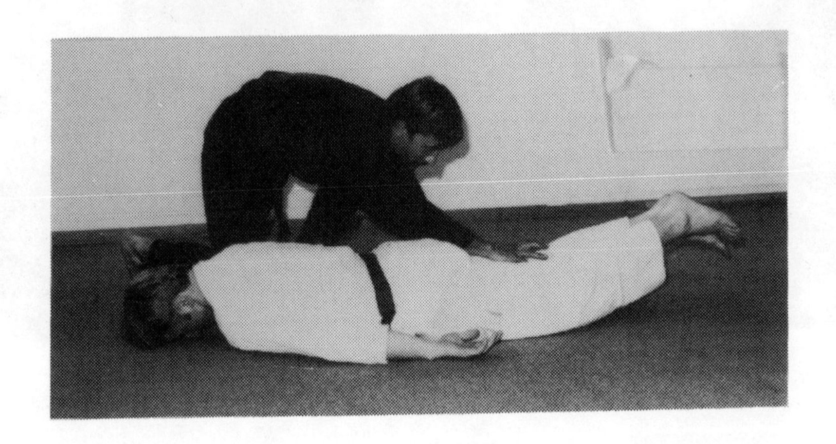

SHOULDER THROW (Yoshi Geri)

Deflect the opponent's attack with the left arm and seize his wrist. Step through with the right leg crossing the opponent's path and place the opponent's arm over your right shoulder. Flex the knees and bend forward at the waist. Simultaneously pull toward your left foot with both hands. This will enable you to "carry" him over your back and right shoulder.

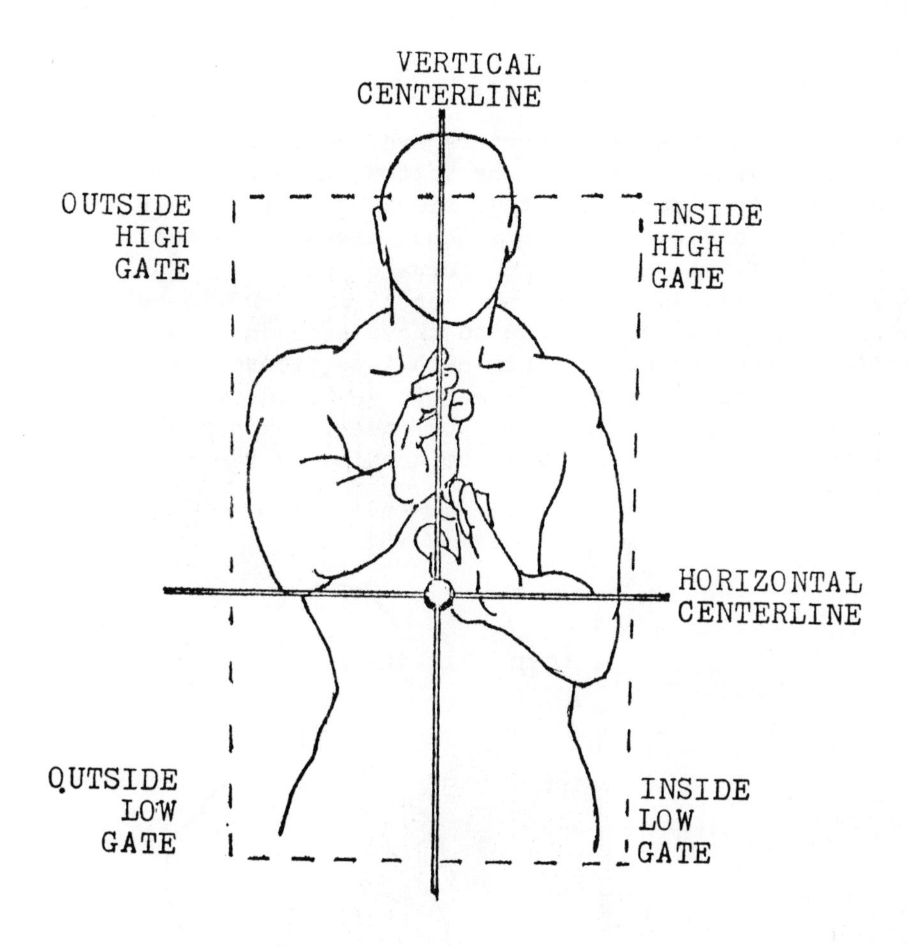

CENTERLINE THEORY---

The Centerline is an integral part of the Wing Chun style. It is the nucleus on which the defenses and attacks are based. The Centerline Theory can be seen in stances, hand positions, shifting of stances and in all advancing and retreating. Facing forward, with the right hand leading, and the left in the center of the chest, the Centerline is defined from the tip of the nose, to the center of the groin, and is touched on both sides by the fingertips. When changing positions, never leave the Centerline unprotected.

THE FOUR CORNERS THEORY---

The boundaries of the Four Corners are the eyebrows at the top; the groin area at the bottom; and the area just past the shoulders on either side. The Four Corner target area is divided into four equal areas, or Gates. The top half of the side of the forward hand is the Outside High Gate. Any attack to this gate will be blocked to the outside. Attacks to the Inside-Gates are deflected to the inside. Within each gate, there are also two separate areas as seen in the side view, a forward area will be blocked by the forward hand. Attacks to the rear area will be handled by the hand that is back.

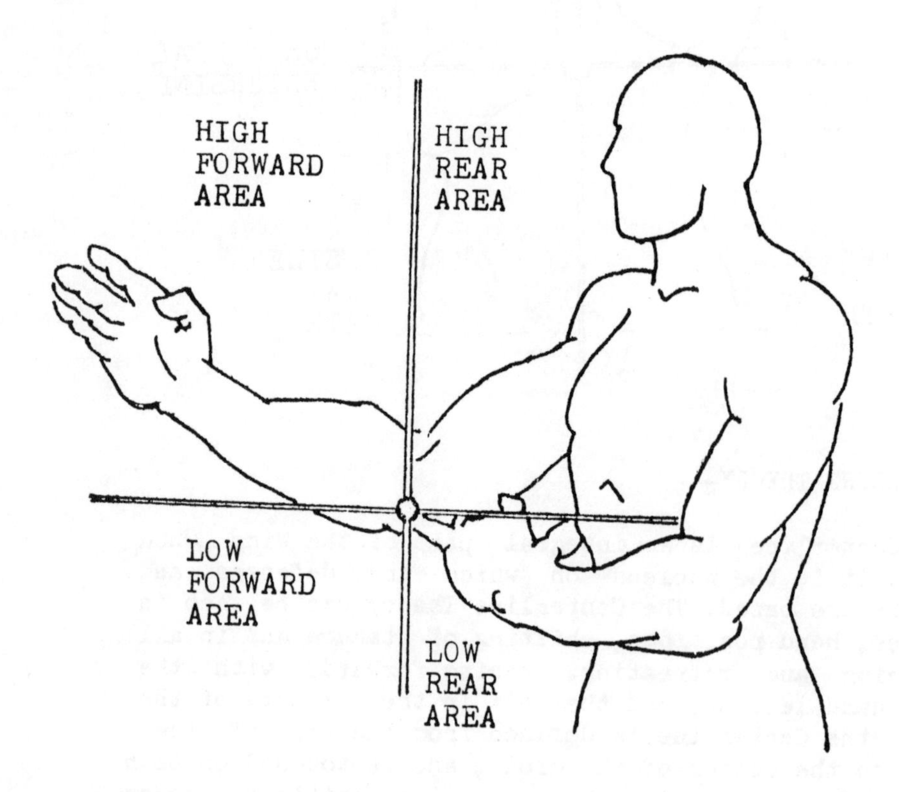

THE PRINCIPLES OF SELF-DEFENSE

DETERMINATION- the master knows, through his own resoluteness, that he will determine the outcome of the aggression and acts accordingly.

CALMNESS- the master's mind remains cool and composed in the face of violence, this allows him to calculate.

CONSERVATION- the master will try to maintain optimum agility and endurance by avoiding movements which are unnecessary.

POWER- the master's goal is to retain his balance at all times, while disrupting that of his opponent.

DIRECTNESS- this is accomplished by keeping the arms and legs as near as possible to the central balance point.

PAIN- "no pain, no gain," is an adage which can be equally applied to practicing as well as defending.

REACTIONS- these must be honed to the sharpest edge possible.

THE SEVEN REQUIREMENTS OF PROFICIENCY

1. Have faith in your selected style of martial art.

2. Respect your chosen master, who has sincerely accepted you as a student.

3. Concentrate during training. There are three teachers in the Dojo: the Sensei; your Eyes; and your Ears.

4. Cultivate patience during training.

5. Practice co-ordination during training.

6. Practice as much as possible.

7. Acquire confidence in yourself through your art.

忍術

The beginnings of Ninjitsu are not clear, but there can be no doubt that it is of Japanese origin, though it was greatly influenced by Chinese military spying techniques. The origin of the word itself stems from a war between Prince Shotoku and Moriya over the land of Omi in the sixth century. During the conflict, a warrior, Otomo-no-Saajin, contributed greatly to the victory of Shotoku by spying out valuable information. For this service, Otomo was awarded the name "Shinobi" which means "stealer in". From this ideogram the character for Ninjitsu was derived.

The role of the Ninja is best used to illustrate the meaning of the art. The Ninja was assigned missions to gain information about the enemy and to sabotage his operations. All Ninja were highly qualified in hand-to hand combat and were required to be proficient with at least three major weapons.

The Ninja were classified into three groups: Jonin, or "upper man" represented the ninja groups and drew contracts with users of the ninja services. His second in command, the Chunin or "middle man", was the group's sub-leader. The Genin, or "lower man" was the agent to whom the actual missions were assigned.

Most famous Ninja were born and trained in the Iga of Koga areas of Japan. Surrounded by mountains and wild areas, the villages where these clans dwelt were almost inaccessible to the casual traveler and enjoyed the protection of nature against their enemies. By virtue of this remoteness and isolation, the Ninja and his trade-Ninjitsu- remained a closely guarded secret.

The Art of Invisibility is NOT some magical technique which enables you to disassemble your body and then re-assemble yourself someplace else; or change the structure of the body so that it remains in view but is transparent. The Art of Invisibility is, the skills you employ to make yourself unseeable, and, in this context, the art itself becomes almost a philosophy.

Inpo, the Art of Hiding, was an integral part of the Ninja skills. He took advantage of every possible object, natural as well as man-made, to conceal himself. In fact, his ability to hide himself so completely was what gave rise to the legends that the Ninja could make themselves invisible at will. Since he was a master of controlled breathing techniques, he was able to remain motionless for long periods. Thus employing the technique of Kagashi-no-jitsu, which states that the eye sees movement first, sees silhouette second, and color third.

Tonpo, the Art of Escaping, was vital to the success of any mission, since capture would undo everthing. And, like every other phase of Ninja activity, nothing was overlooked that might assist him. The technique of Kasumi, or "hazing" to temporarily blind an enemy was included in the ruses of the Ninja. Moonlight created an endless variety of shadows in which the Ninja could hide during an escape.

The Classical Exercise of Invisibility, however, is the **Mi Lu Kata**, or Lost Track Form, a portion of which is shown here.

Uke stands in the basic Sesgae, or At Ease, position. Nage comes up behind him and taps him lightly on the left shoulder. This technique takes its name from the concept that no one can see both sides of a coin at once. Making sure that the enemy detected his actions, the Ninja would pretend to aim at the rear gate of a heavily guarded castle. While drawing attention to the back, he slipped through the front. A similar trick was to design two simultaneous actions, one as a feint, and the other the real intent.

In Figure Two, Uke looks to his left, anticipating that whoever touched him will be standing on that side. Nage moves quickly down and to the right side of Uke, remaining outside his peripheral vision. At this moment, Nage is totally invisible to Uke. Note that Nage focuses his attention on the base of Uke's skull. This enables him to anticipate the next movement.

In Figure Three, Uke suspects that Nage is behind him and begins to turn to his right in an effort to catch him. Nage moves quickly and silently to Uke's left, drawing his right foot in and "closing his stance". This eliminates the possibility that Uke, in turning, will glimpse his right foot. Nage also raises his hands in defense, but does not touch Uke. He continues to focus on the base of the skull.

Uke continues to turn to his right. Nage steps quickly forward with his right foot, turning back-to-back with Uke and executing the first half of the Mi Pu, or Invisibility Step. The pivot is so named because for a split second, Nage must turn his back to Uke, thus rendering both the attacker and the defender unseeable. Nage must "point" on Uke's skull while turning to insure that he turn as quickly as possible.

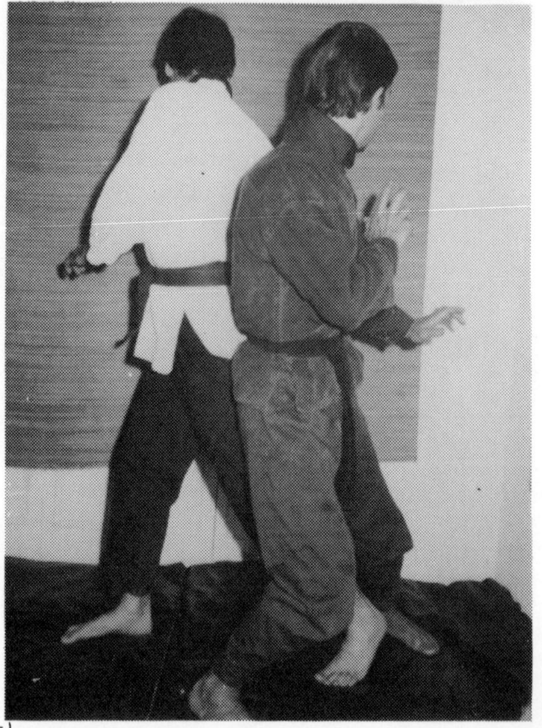

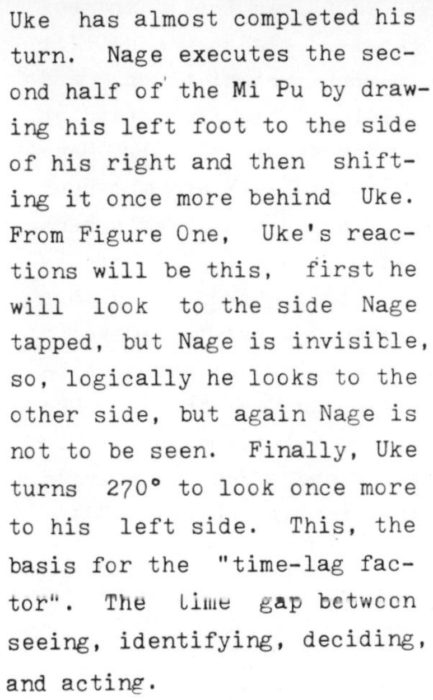

Uke has almost completed his turn. Nage executes the second half of the Mi Pu by drawing his left foot to the side of his right and then shifting it once more behind Uke. From Figure One, Uke's reactions will be this, first he will look to the side Nage tapped, but Nage is invisible, so, logically he looks to the other side, but again Nage is not to be seen. Finally, Uke turns 270° to look once more to his left side. This, the basis for the "time-lag factor". The time gap between seeing, identifying, deciding, and acting.

Figure Six illustrates the Nyudaki-no-jitsu or Sentryhold. As Uke steps forward with his left foot, Nage whips his left around to strike Uke in the throat with the ridge of his wrist. Simultaneously, he hits Uke in the right kidney, thus breaking his balance backward. The term Nyudaki-no-jitsu refers to discovering a guard's weaknesses or shortcomings in order to penetrate the enemy camp.

The throat strike causes Uke to draw in his breath sharply, preventing him from making an outcry. Nage presses his advantage by sliding his left a m around Uke's neck in preparation for the classical Japanese Strangle. Uke tries to defend by grasping Nage's forearm. This is to no avail, since the throat attack will have damaged the pharangeal nerves which control the diaphragm, making it impossible for Uke to breathe.

Nage applies the Japanese Strangle, placing his left palm on his right bicep, and the palm of his right hand on the back of Uke's skull. Nage relaxes his left arm and uses his right to press Uke's neck forward into the "V" of his left elbow. Pressure is thus directed against the Carotid Artery, which collapses, ribbon-like as Uke inhales. Uke will become unconcious within five seconds due to oxygen starvation of the brain.

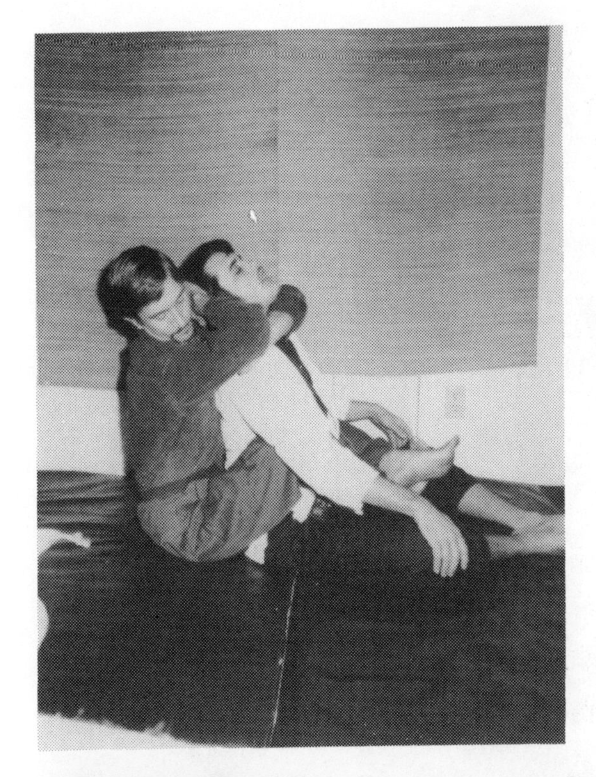

Nage sits down on his right haunch, lowering Uke quietly to the mat. Nage wraps his left leg around Uke's waist. Next, he entwines Uke from the right side with his right leg. Locking his ankles, right on top, Nage squeezes Uke's stomach, forcing the air out of Uke's lungs and abdomen. This insures that Uke will lapse into unconciousness, and prevents him from making any sound as he falls.

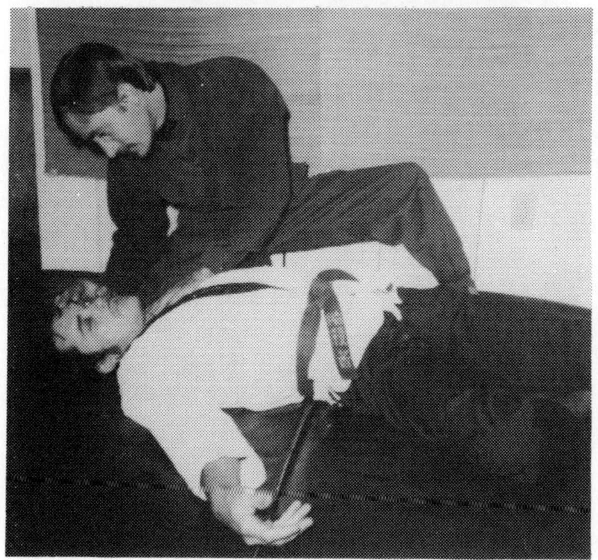

In Figure Ten, Uke lies unconcious. Nage kneels at his side and checks to determine Uke's level of unconciousness. He does this by noting the pupillary reflex (if the pupil is dialated, insufficient blood is reaching the brain), the rate at which the pupil reacts to light indicates how restricted the blood flow is. He also checks the Carotid Pulse, on the side of the neck next to the trachea, with the fingers of his left hand. He will check both sides.

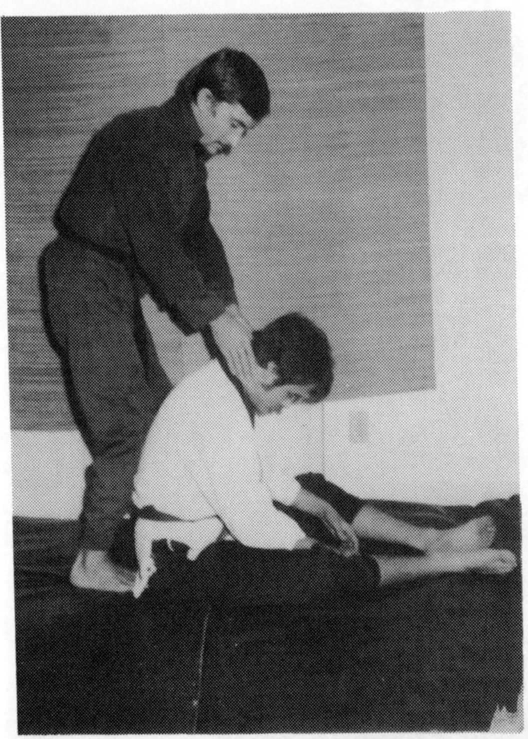

Nage raises Uke to a sitting position and props him against his right knee. Placing his thumbs at the base of the skull, Nage massages the neck upward with his fingertips to restore circulation. During this process, Nage also checks the cricoid cartilage of the larynx for damage, and strokes the carotid sheath, insuring that no large blood clots have formed. Since the artery is not ruptured by this attack, no bruising should be seen.

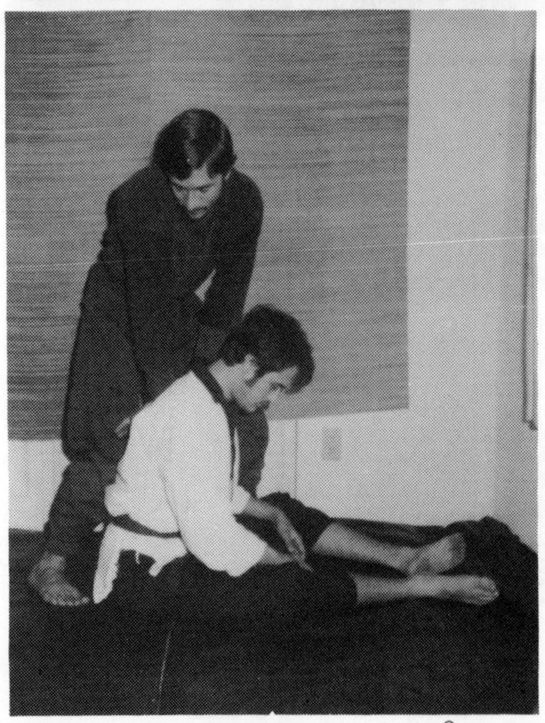

Shifting to Uke's left side, Nage supports him by placing his left hand on Uke's left shoulder. Striking directly downward, Nage hits Uke between the shoulder-blades on the seventh thoracic vertebrae. This stimulates nervous action. Occassionally, Uke's epiglottis will fall shut, sealing the windpipe. This action jars the body sufficiently to clear the airway. Respiration should begin spontaneously.

REGARDING THE USE OF STRANGLE-HOLDS

There are two types of strangle-holds: 1) Using the coat collar, or 2) Using the arms or legs around the neck. Neck holds can be used alone or in combination with other techniques. They involve pressing or squeezing against the throat or carotid artery. The result is to stop breathing, or the blood supply to the head. This makes the opponent dizzy or unconcious, putting an end to his resisting power and rendering him temporarily out of action.

HOWEVER, WHENEVER IT IS NECESSARY TO USE A NECK HOLD, BE EXCEEDINGLY CAREFUL. THE MAN WHO HAS BEEN RENDERED UNCONCIOUS BY A NECK HOLD WILL NOT REVIVE AUTOMATICALLY! HE MUST BE BROUGHT BACK BY EXPERT RESUSCITATION.

Neck holds are usually applied when the opponent starts to rise after a throw. **It is most important to act** quickly, before he can recover himself. There are several positions in which an opponent might fall as the result of a throw, hence the various neck holds which have evolved in the martial arts. Neck holds are so simple that women, and even children, can use them easily against more powerful opponents.

THE WORLD'S DEADLIEST FIGHTING SECRETS

The Dance of Death is itself the most deadly defense form possible. Each movement is smooth, flexible, and flowing from one technique to another. This defense can be used against the hook, roundhouse, cross, and with slight variation, the jab. An effective time for execution of the form is given and the complete form should be performed in no less than the time stated. It can be effectively executed in less time, but a median time is given. Keep in mind the speed and flow of movement in the performance of the form.

KATA DAN'TE
Effective Execution Time :05 seconds

The Dance of Death is itself the most deadly defense form possible. Each movement is smooth, flexible, and flowing from one technique to another. This defense can be used against the hook, roundhouse, cross, and with slight variation, the jab. An effective time for execution of the form is given and the complete form should be performed in no less than the time stated. It can be effectively executed in less time, but a median time is given. Keep in mind the speed and flow of movement in the performance of the form.

Here we see the attacker in position to throw a right hand punch, while the defender is at the ready.

The attacker has stepped in with his punch. The defender immediately moves inside of the attack catching it at its beginning. His left, closed hand, deflects and checks the strike at the same time that his right hand effects the counter-attack. Note the strong definite stance of the defender. His right hand performs a series of attacks with its first motion: The Palm Heel strikes the chin, snapping the head back while the strike drives upward, splitting the upper lip and crushing the teeth, and here in a continuous motion, the hand goes upward crushing the nasal cartilage and bone, while tearing the nose upward also, and away from the face. This combination block-and-counter strike is performed in one strong continuous motion.

After completing the upward motion the defender brings his fingertips and nails down across the eyelids, tearing them open, and clawing the eye itself. A tearing or slashing motion will do much more damage here than an individual jab.

Immediately following the right hand attack to the eyes, throw a left hand open palm slap to the side of the face. After impact, but in the same circular motion, grip the cheek and snap-tear it off. (Tiger Claw)

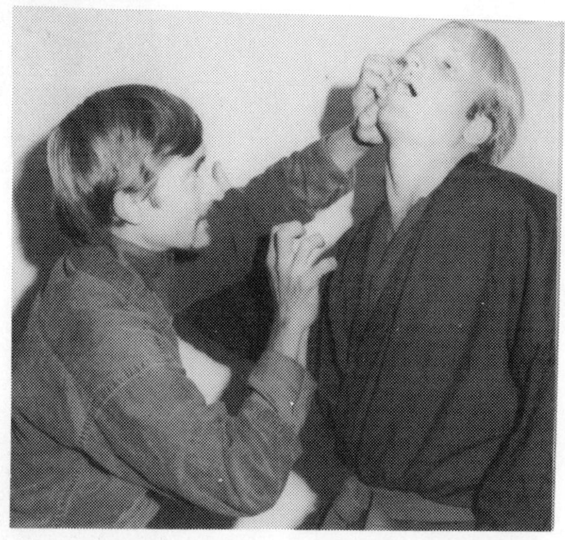

Execute the Tiger Claw strike using the right-hand. The beginning of the grab-tear is shown.

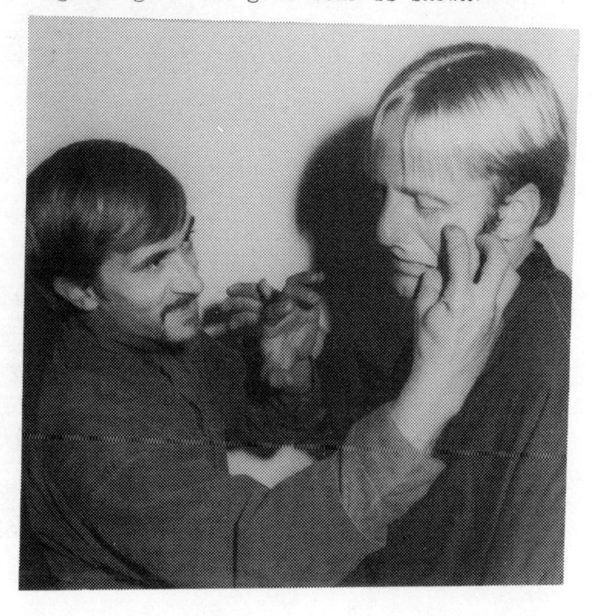

Retract the right hand from the face, and throw a short, thrusting, forceful open hand hook to the throat (Tiger Mouth Strike). Hit as hard as you can and drive the hand forward as far as it will go into the throat.

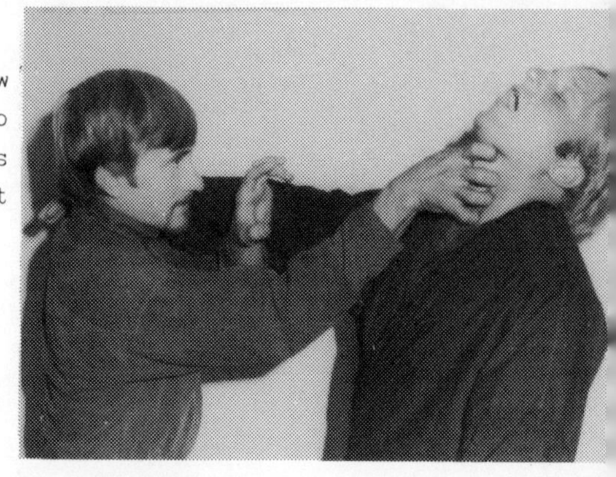

Completing the tear and maintaining the downward circular motion, release the throat and continue the circle up and under the groin with an open hand slap to the testes. (Monkey Stealing Peach) After the slap, grab and de-groin in an upward and backward movement (toward you). Note how the left hand locks the attacker's arm, and helps maintain control over him.

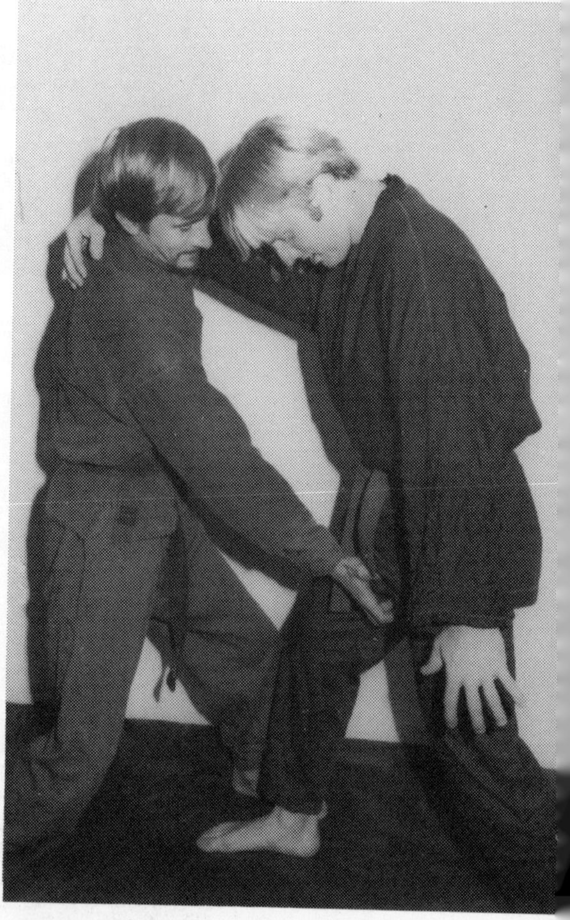

ter the right hand de-groining technique, ur closed right hand should be at your ght side. From here throw a right elbow, ilizing the arm to push him into the blow lbow Stroke).

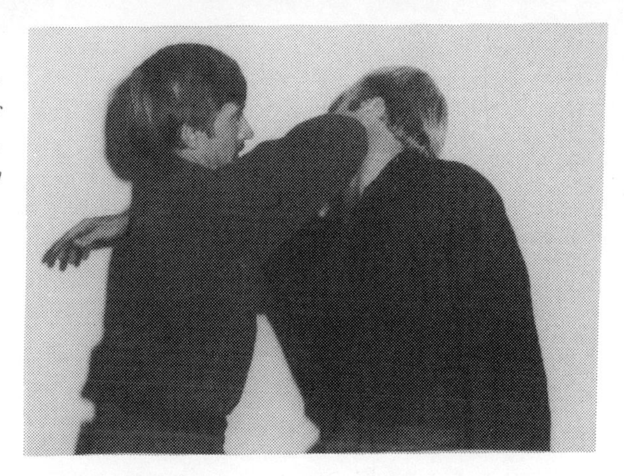

At this point use your left arm-lock to twist the attacker forward and down into a right knee strike to the face and a right elbow strike to the base of the skull. Maintain control with the arm-lock.

From here, using the attacker's arm to pivot, turn 180° bringing him along, controlling his movement by your motion and the arm - lock. This movement should be quickly executed. It is simple, as we are slamming him to the mats in a circular motion, along his weak lines of balance. As you turn, drop to your left knee, and the right leg which was used for the bent knee strike will come down with a circular motion on its sole with the knee bent. As the attacker hits the ground, the force of the impact along with the pressure from your controlling left hand and arm is used to break the arm at the shoulder joint as shown. The attacker is pinned to the ground by the control of the arm. Your right hand breaks his arm at the elbow with a closed-hand strike, (Hammer Fist) and then continues downward to strike the base of the neck with the Sword-Hand Technique (Shuto).

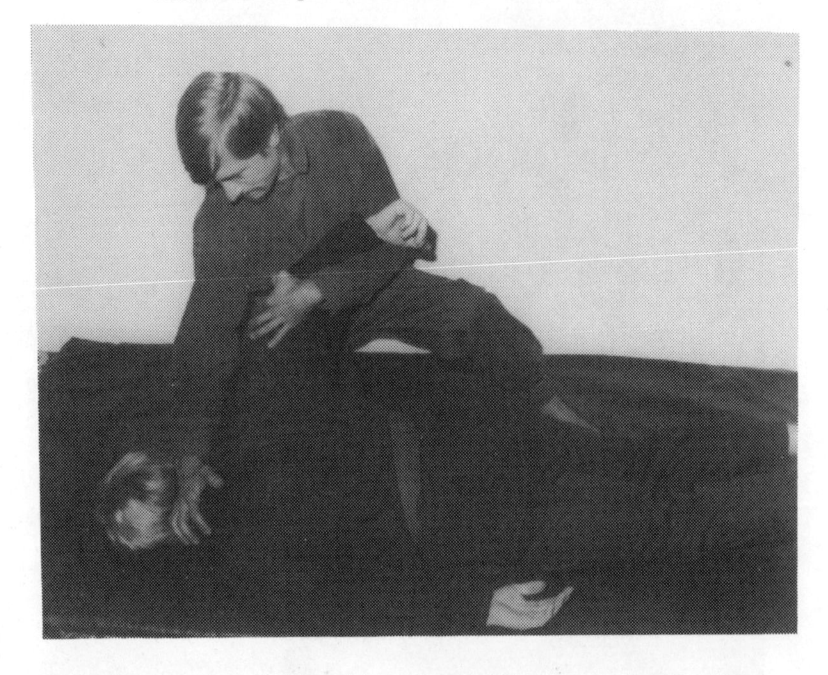

After the break-and-strike, release the attacker and leap up, coming down with both feet on your assailant. Your right foot should land on his head or neck, while your left foot lands on the base of his spine. The combination of the downward crushing motion and the stretching action on the spine as the feet are spread, effectively snaps the backbone (Dragon Stamp).

Reach down and while lifting the head and tearing the face from the rear, snap the head to the side, breaking the neck. After the break, smash the face or side of the head to the ground, putting the entire body weight and strength into the smash. The left knee breaks the attacker's spine.

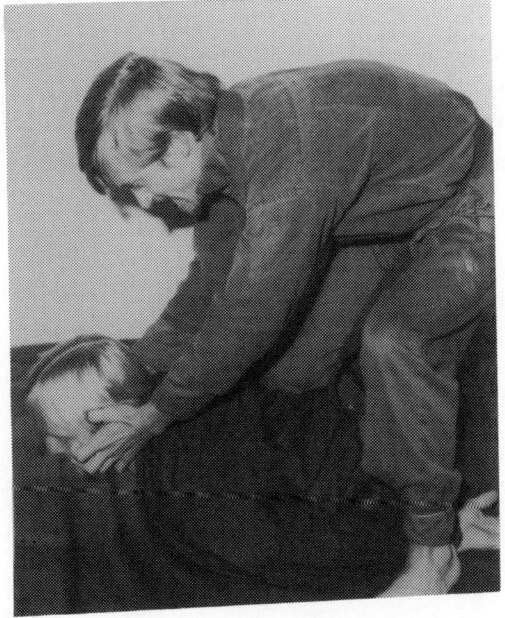

Now with the left hand, grab his left side and pull with a circular, counter-clockwise motion at the same time, lifting with the right hand, and stepping to the left with the left foot to roll him over.

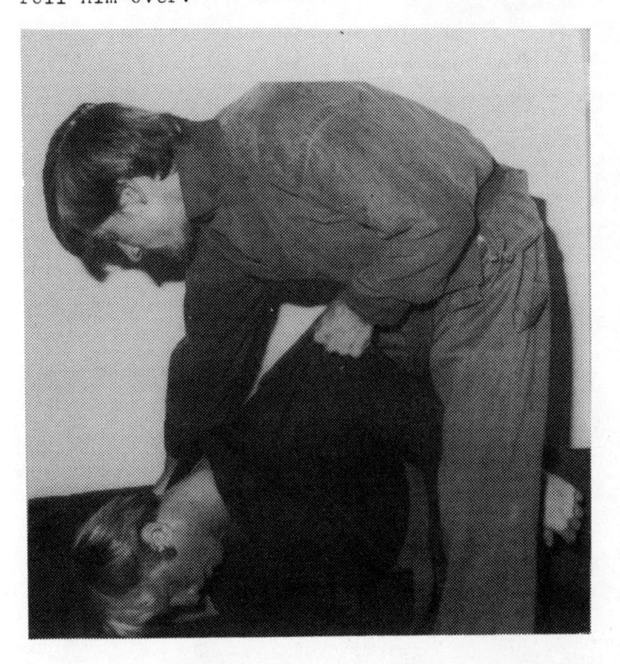

After completing the turn, leap up, coming down with a crushing stomp once again. The right foot should land on the throat or face, while the left lands on the groin or lower abdomen.

After the double stomp, reach down with a two-hand forward cross tear, ripping the mouth and lips open as far back as the ears. Grip the ears and remaining flesh with a reverse and inward cross-tear, and draw everything to you.

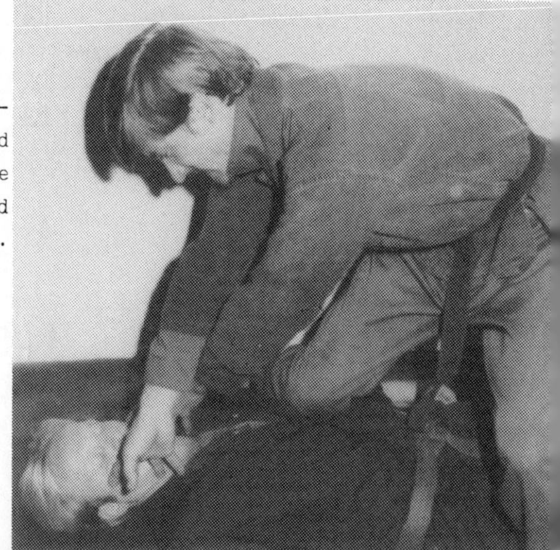

With a quick raise and drop with your right knee (using your full weight) crush the head to the ground. After the knee-drop, cross your right leg over his body to a cross-legged stance, always keeping the hands in a defensive position, and move away.

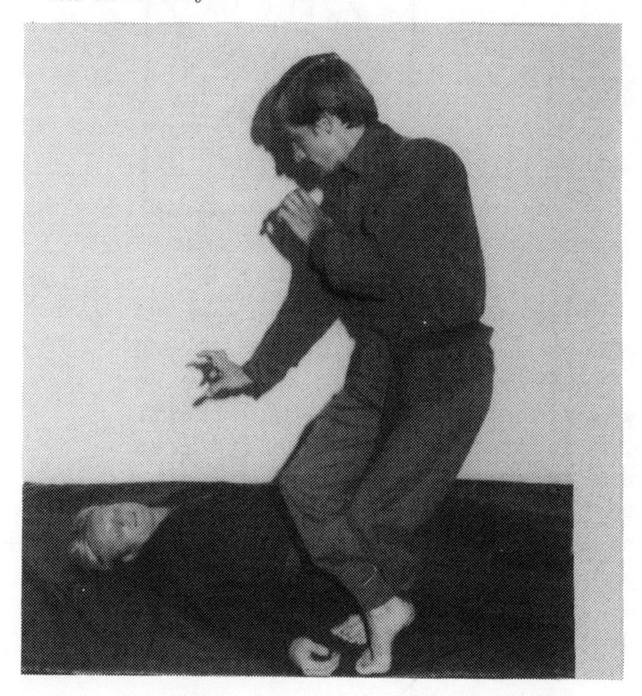

DIM MAK
"The Death Touch"

This is the fist, the weapon, the tool. It must be clinched tightly. Now you have your weapon. How do you use it? We use it in two ways. First, in the more or less ordinary way, as an extension of the body, specifically, the arm. But there is another way. The fist can be submerged into the arm in such a way that it is only the point of a spear, so to speak. Here the arm, not the fist, is the weapon. In this type of attack the arm is rigidly straight, and not bent at the elbow. The power generated in this long attack is truly enormous but, as you can well guess, this must be meshed with a short attack or it is fairly easy to defend against.

When you strike it must not be haphazard. Every hit must have a target, and every target, a reason. Generally, the temples, throat, solar plexus, and groin are the best targets, for an effective hit in any of these is often fatal.

I do not stress kicking. There are many good methods, but there are only twenty-four hours in a day. If I could kick expertly, I would be an imperfect boxer, and in all humility, I tell you that my boxing is not imperfect.

DIM MAK---
The Delayed Death Touch

At specific times of the day, certain vital areas of the body are especially vunerable to attack. This is because the blood comes close to the surface at different times of the day. One need only be aware of the course of this circulation, and attack the appropriate vital point, at the time the blood is near the surface. Thus, injury is certain and death probable. Proper co-ordination, concentration of mind and body, correct method of striking, breath support, transmission of Chi, and a thorough knowledge of the vital and fatal points as they pertain to the timetable combine to make the touch effective.

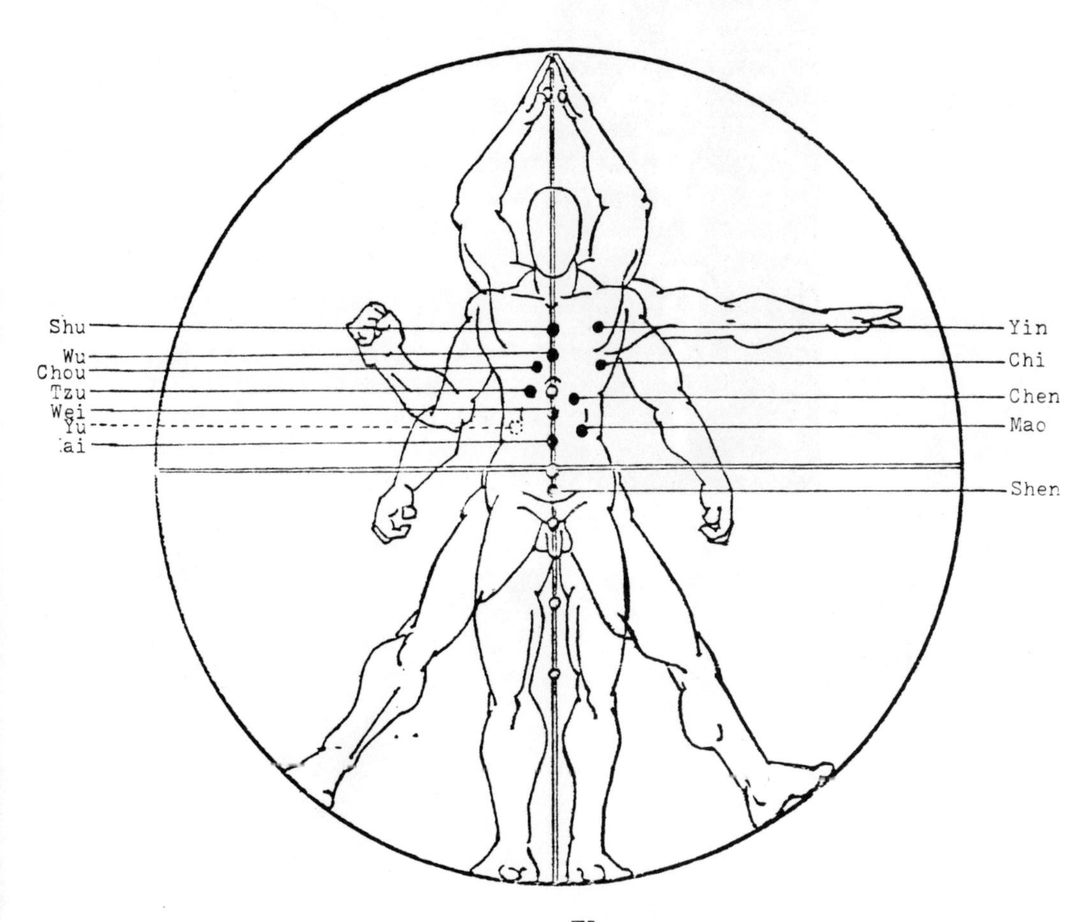

東宝
SAMURAI

SAMURAI CREED

I have no parents; I make the Heavens and the Earth
my parents.
I have no home; I make Tan T'ien my home.
I have no divine power; I make honesty my Divine
Power.
I have no means; I make Docility my means.
I have no magic power; I make personality my Magic
Power.
I have neither life nor death; I make A Um my
Life and Death.

I have no body; I make Stoicism my Body.
I have no eyes; I make the Flash of Lightning
my eyes.
I have no ears; I make Sensibility my Ears.
I have no limbs; I make Prompitude my Limbs.
I have no laws; I make Self-Protection my Laws.
I have no strategy; I make the Right to Kill and
to Restore Life my Strategy.
I have no designs; I make Seizing Opportunity by
the Forelock my Designs.
I have no miracles; I make Righteous Laws my Miracle.
I have no principles; I make Adaptibility to all
Circumstances my Principles.
I have no tactics; I make Emptiness and Fullness
my Tactics.

I have no talent; I make Ready Wit my Talent.
I have no friends; I make my Mind my Friend.
I have no enemy; I make Incautiousness my Enemy.
I have no armor; I make Benevolence my Armor.
I have no castle; I make Immovable Mind my Castle.
I have no sword; I make No Mind my Sword.

SWORDSMANSHIP---

Though not the oldest weapon in Japan, the sword is the central weapon on the martial hierarchy. It can rightfully be called the "soul weapon" of the warrior class. The Two chief arts of the sword encompassed by the budoka are known as Kenjitcu and Iai-Jitsu. Through these systems the budoka as a whole can be understood and appreciated. The art of the sword has literally carved the nation's policy.

The earliest Japanese swords were made of wood and stone. It was not until the coming of metal to Japan around the second century B.C. that a true blade was formed. Early military leaders placed a high priority on sword design and tactics of employment, but it was during the Nara Period (710-794) that the single-edge, curved blade, two-handed sword came to prominence.

The bushi, armed with the sword, became the important social class in spite of efforts of court nobles to restrict thier rise. By the early twelfth century, the court lost its position to the military families, who had increased their stature and backed their aims with well armed troops.

Use and upkeep of the bushi's sword was defined by a strict and minutely detailed etiquette. The bushi bore such customs quite proudly, and suffered the waiver of them to no one. In front of a person of unknown intention, the bushi was always careful to keep his long sword close at hand. If he positioned it at his right, as he knelt in respect, it signified his intention was not hostile, because in that position, the sword could not easily be brought into play. However, if it was on his left side, the host knew that the caller was either reluctant to accept his hospitality, or that the guest had evil intentions.

At a friend's house, the bushi might remove his long sword in the outer hall and leave it on the sword rack which was provided there, or turn it over to a servant, who had been carefully schooled to treat the blade of nobility with the fullest respect and care; it would be carried by means of a silk cloth.

The sword was all things to the bushi, a divine symbol, a manly weapon, a badge of office and honor, which signified his noble ancestry. The sword became his "soul" vested in him by tradition, which he did the utmost to preserve.

The dimensions of the sword varied with individual tastes, but some geralizations are possible. Blades were approximately two feet long for the long sword, and about one foot long for the short sword. They were one and a quarter inches in width, the back of the blade was one quarter inch thick and tapered to a razor edge. At the back of the blade and along its length was sometimes found a "blood groove" that served to make withdrawal from the enemy's body easier and which also served to collect the enemy's blood to make the blade easier to clean.

The handguard was functional and often ornamental. Miyamoto Musashi, the most famous swordsman of feudal Japan was said to have inscribed both faces of his guard with the legend, "the sword which takes life" on the side toward the enemy, and the "sword which gives life" on his side.

SWORD TYPES---

DAITO- or Long Sword, was a single-bladed curved sword, worn thrust through the sash or belt of the warrior, on the left front side of the body, positioned with the cutting edge upward. The Daito is drawn in a "sky to ground" motion, indicating that as the blade cleared the scabbard, the point describes an arc which brings the edge in and downward toward the opponent's head.

SHUTO- or Short Sword, is by far the more wicked of the two swords commonly used. It is not even studied by Kendoists until reaching the third Dan or Rank. It is generally worn in front or on the right side with the edge downward, and is drawn with the left hand. The use of the two swords simultaneously is known as Ni-To-Ryu, and requires many years of practice. The basic "en garde" position is with one weapon in each hand, with the tips overlapping by no more than four inches.

NINJA-TO- or Sword of Darkness, is a third type of sword found in Kenjitsu, and is the product of a fertile imagination, being quite functional. Unlike the Daito, or the Shuto, which may have been decorated in numerous ways to the satisfactions of the owner, the Ninja-To is totally devoid of any decorative design. Further, it serves not only as a weapon, but also a tool of its master. First, the scabbard is longer than the blade and could be used to concealing objects, and in conjunction with the sword as a club or mace. The sheath was open at both ends, allowing it to be employed as a blow-gun, or a snorkel-like breathing device. The tsubo is larger than the conventional handguard, and generally rectangular in shape. This served as a foothold if the sword were leaned against a wall. Once the summit was reached, of course, the sword would be drawn up to its owner by means of the attached cord. This cord also served as a tourniquet in the event of an injury.

Naturally this does not exhaust the types of sword found in the Orient. Notably absent is the Broadsword. But this is primarily of Chinese origin and is used in a one-handed manner. Likewise, there are numerous smaller bladed weapons, dirks, daggers, shuriken, and so on, but these too have been omitted, ince they are not used in the two-handed manner.

The techniques which follow apply equally to all of the three sword types listed herein. The terminology, as well as the target areas, are identical. The primary difference in the Ninja-To as compared to the Daito and Shuto, is its emphasis on concealment of the blade and the numerous additional uses of the scabbard as an auxilary tool.

CHAMBERED STANCE

Draw the left foot to a position near the right instep. Place the sword hilt on the right hip, the haft is now held in a "baseball-grip". In performing this action, the Ninja takes himself out of range of the enemy's attack and "loads" his own attack simultaneously. From this stance, it is possible to move into any of the three basic parrying positions, but it is as the preparatory movement for the "Straight Thrust", that it is most effective.

THE STRAIGHT THRUST

From the Chambered Stance, step quickly forward with the left foot, advancing along a direct line of attack to the Centerline. The point of the sword describes an arc from the right shoulder, over the line of the eyebrows, directly toward the solar plexus of the enemy. Added momentum is given to the thrust by the forward movement of the body. Note that the sword enters the body at a slight downward angle. This prevents the point from skipping along the ribcage. The right hand is the driving force of the thrust.

JODAN-CHIEN

The Upper Quarter Sword Block. From the Draw, drop the left elbow to the left hip, and swing the blade to the outside line. This brings the hilt to a position horizontal and about ten inches in front of the Hara. Grasp the lower end of the hilt with the right hand and "sweep" the area from Hara to forehead with the edge of the sword. This action will close the High-Gate to the enemy's attack.

CHUDAN-CHIEN

The Middle Quarter Block. From Jodan-Chien, swing the sword across the body, sweeping the area in front of the chest with the sword edge. This closes the middle level to the incoming attack of the enemy, and deflects that attack to the outside line. Added impetus is given to the stroke by twisting the hips as the sword crosses the body. Note that this parry is most effective against a Straight Thrust.

GEDAN-CHIEN

The Lower Quarter Block. From Chudan - Chien, strike downward to the Inside Low Gate, sweeping the area from Solar Plexus to knee with the sword edge. This powerful block may also be used as a strike against the enemy's low line. Note that in this, the wrists become crossed The right wrist is the actual motive force for this parry, while the left acts to hold the hilt "on line" for the deflection.

THE KOGA-NINJITSU DRAW

Note that this cut is made "ground-to-sky", that is to say, on the "upswing". The attack is directed against the arm, leg, or groin. Beginning in the basic fighting stance, step forward with the left foot, grasping the sword hilt with the left hand and the sword scabbard with the right. Whip the sword out of the scabbard in a single, powerful motion. Focus the attention on the Third Eye of the enemy and "fix his gaze".

Terry Stephen Duncan
3rd Dan, Shotokan

In Hsiao Chien Do, you handle a very sharp blade. It can slice easily and do great injury. Thus, the utmost in concentration is required. It is the combination of physical and mental ability which leads one toward the goal. Alone, the warrior prepares himself for the Inner Battle by taking three deep breaths.

The Sword is drawn from the scabbard in a single, strong continuous motion.

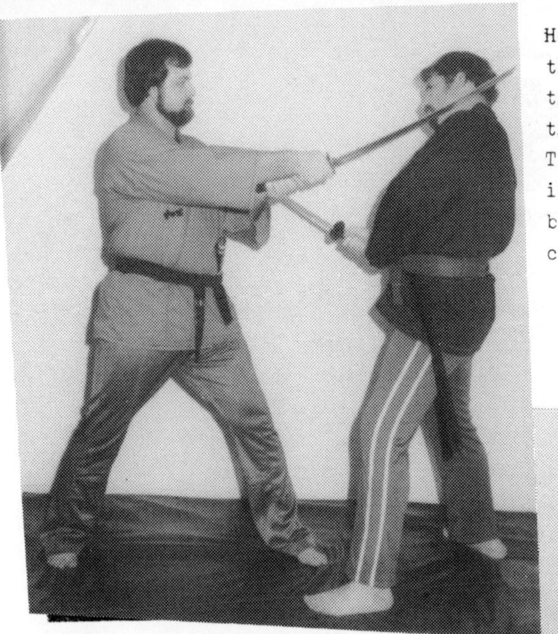

He raises the Sword in preparation for the "stroke", which slices down toward the left side of the neck and penetrates to the mid-line of the body. The stroke is made with the last six inches of the blade, and is performed by drawing the Sword toward the chest, cutting on the backstroke.

He is now in position to complete the form by flicking away the blood which covers his blade. Called "Chiburi" by Kenjitsuka, this act is performed by snapping the wrist down and to one side of the original line of attack.

In the final action, the Sword is drawn across the scabbard mouth, guided by the left thumb and forefinger. Preparing it for return to the scabbard, the point is inserted and the blade driven home by the right hand in a single snapping motion.

CHRISTOPHER HUNTER

The word Ninjitsu has been translated as the "Silent Way"; the followers of that way are known as Ninja, the invisible assassins. Looked for they cannot be seen, listened for they cannot be heard.

Espionage was their trade in ancient times, a trade they referred to as the Great Game, and at which they had no equal. They were primarily small men with broad shoulders, who believed that nothing is impossible.

We make availible here for the first time a glimpse into the closed society of the Ninja. The secrets revealed are not found in other books on the martial arts, neither are they presented lightly to the general public. Their publication is intended only as the study of an ancient and obscure art form.

We have no fear that the techniques illustrated will be misused, since their mastery requires years of persistent practice.

BIBLIOGRAPHY

ASIAN FIGHTING ARTS, Donn Draeger; Berkley Publishing, 1974

GRAY'S ANATOMY; Running Press, Philadelphia, Pa. 1901

HAPKIDO, Bong Soo Han; Ohara Publications, Los Angeles California, 1976

HANDBOOK OF JUDO, - ene Lebell-L.C. Coughran; Cornerstone Library, 1972

HINDU BOOK OF DEATH, Dr. L.W. DeLaurence, Benares India Publishing Co., 1905

ILLUSTRATED KODKAN JUDO; Kodokan, Tokyo, Japan, 1955

NINJA, Andrew Adams; Ohara Publications, 1972

SOLDIER'S BASIC TRAINING HANDBOOK, Headquarters, Department of the Army, PAM 21-13, 1969

SECRET FIGHTING ARTS OF THE WORLD, John F. Gilbey, Charles Tuttle Publishing, Rutland, Vt. 1963

WING CHUN KUNG FU, J. Yimm Lee; Ohara Publications, 1972

WORLD'S DEADLIEST FIGHTING SECRETS, John Keehan; Black Dragon Fighting Society, 1968

POSTSCRIPT:

Now that you have read and understood the information contained herein, it is advised by the author that you return to the beginning and read through twice more. In this way the Secrets of the Empty Hand will be placed at the subconcious level of Mind, ready to be recalled should the need arise...